Ana Beatriz Pereira de Castro
Lázaro Costa de Souza
Adna Cristina Barbosa de Sousa

ALTERNATIVE AND BIOLOGICAL CONTROL OF PULLOIDS (Rhopalosiphum spp.)

Ana Beatriz Pereira de Castro
Lázaro Costa de Souza
Adna Cristina Barbosa de Sousa

ALTERNATIVE AND BIOLOGICAL CONTROL OF PULLOIDS (Rhopalosiphum spp.)

Alternative pesticides and entomopathogenic fungi: sustainable agents for controlling aphids(Rhopalosiphum spp.)

ScienciaScripts

Imprint

Cover image: www.ingimage.com

This book is a translation from the original published under ISBN 978-613-9-68874-6.

Publisher:
Sciencia Scripts
is a trademark of
Dodo Books Indian Ocean Ltd. and OmniScriptum S.R.L publishing group

120 High Road, East Finchley, London, N2 9ED, United Kingdom
Str. Armeneasca 28/1, office 1, Chisinau MD-2012, Republic of Moldova, Europe
Printed at: see last page
ISBN: 978-620-7-45878-3

ALTERNATIVE AND BIOLOGICAL CONTROL OF APHIDS (RHOPALOSIPHUM SPP.)

ALTERNATIVE PESTICIDES AND ENTOMOPATHOGENIC FUNGI: SUSTAINABLE AGENTS FOR CONTROLLING APHIDS (RHOPALOSIPHUM SPP.)

***ANA BEATRIZ PEREIRA DE CASTRO*[1]**

FEDERAL UNIVERSITY OF PARAÍBA CAMPUS I/BIOTECHNOLOGY CENTRE -

CBIOTEC)

***MR LÁZARO COSTA DE SOUZA*[2]**

PARAIBA RESEARCH, RURAL EXTENSION AND LAND REGULARISATION COMPANY - EMPAER

***DR*[A]*. ADNA CRISTINA BARBOSA DE SOUSA*[3]**

PROFESSOR AT THE FEDERAL UNIVERSITY OF PARAÍBA CAMPUS I/CENTRE OF

BIOTECHNOLOGY - CBIOTEC E-MAIL

CONTACTADNACSOUSA@GMAIL.COM

JOÃO PESSOA - PARAÍBA OCTOBER - 2024

ACKNOWLEDGEMENTS

Above all, I thank God for his graces in my life, love and faithfulness. Your self-giving on the cross inspires me to be charitable and makes me constantly seek holiness. May I live according to your plans and your will through the intercession of your Blessed Mother.

I would like to thank my family, especially my father Alexandre Cézar and my mother Josicleide Castro, for being my greatest refuge and support, for encouraging me to be happy no matter what course I chose and making it possible for me to chase my dreams, I would be nothing without you.

To my grandparents, Maria das Neves, Maria José and José Calazans, for being constant examples in my life and inspiring me to be better every day. You will always be in my thoughts and prayers.

To Alex, my big brother, and my sister Alessandra, my best friend and confidante. The greatest gift my parents could have given me was you, it's a great pleasure to share my life and my parents with you.

To my dear boyfriend, Murilo Stabili, for his affection, extreme patience, zeal and romanticism, for the grace of sharing my life with yours and seeking to serve God together.

To my teacher and supervisor Adna Cristina, I express my enormous gratitude for the opportunity, trust and help. Thank you for always being so attentive, welcoming and patient! My degree would not have been the same without your guidance and support in my life.To Professor Ulrich Vasconcelos for the chance to be a tutor in Microbiology I and Microbiology II.

II and gave me the opportunity to experience the marvellous experience that is teaching, especially in the area of Microbiology and Mycology.

To my dearest friends, Geovana Ferreira, Karine Miranda, Pedro Henricke, Maria Luiza, Lucas Medeiros, Rian Fernandes, Bruno Azevedo, Andréa Coelho, Anderson Furtado, Débora Luise and Luís Felipe, who allowed this degree to become lighter every class, lunch and shared despair provided wonderful moments that will remain in my mind forever, your friendship is very precious to me.To the SIC Confirmation Ministry, for bringing me closer to the most beautiful and pure love I have ever known: God's love. Serving with all of you is a great blessing and I know that I have you in my heart.Shrine of the Immaculate Conception a family.I would like to express my sincere gratitude to all those who were directly or indirectly responsible and collaborated so that I could carry out this work.

"Before you were formed in the womb, I knew you; before you were born, I knew you.
had consecrated. (Jeremiah 1,5)

SUMMARY

Aphids (Rhopalosiphum spp.) are very common pests in agriculture and are responsible for transmitting viruses, yellowing and leaf fall, which can lead to the total loss of a crop. Therefore, integrated management with entomopathogenic fungi and natural pesticides has emerged as a sustainable alternative that is less harmful to the environment when compared to agrochemicals. In view of the above, the aim of this study was to evaluate the in vitro lethal action of eleven alternative pesticides and four species of entomopathogenic fungi (Beauveria bassiana, Beauveria brongniartii, Metarhizium anisiopliae and Metarhizium flavoviride) against aphids in order to offer a sustainable alternative for controlling these insects. To analyse the effect of fungi and natural pesticides, bioassays were carried out on n= 40 insects with four replicates (TC - control treatment: distilled water, Treatments 1, 2 and 3: T1, T2 and T3 with natural insecticides and entomopathogenic fungi). The total evaluation time was 20 days. In the bioassay with the entomopathogenic fungi, a confirmed mortality rate was obtained for Beauveria bassiana and B. brongniartii of 96.4% and 100% in a TL_{50} of 5 to 6 days, respectively. Aphids infected with Metarhizium spp. showed no confirmed mortality. With regard to the results obtained from the use of natural syrups, all provided average mortality of 80% to 100%. The most promising formulations were Manipueira and coconut soap, which caused 100% of the aphid population to die in just 24 hours. The results obtained with the use of syrups and entomopathogenic fungi to control aphids show that they are effective and can be used to control this insect and may result in more ecological pest management that has less impact on nature.

Keywords: Pest control. Sustainability. Mycoinsecticides.

SUMMARY

1. INTRODUCTION

Aphids are insects that are considered agricultural pests due to their distribution, their complex life cycle and the direct and indirect damage they cause to plants (Margaritopoulos, 2020). Direct damage occurs through sap suction, which generates consequences such as deformation of plant tissues, disturbances in plant metabolism and changes in the amount of various compounds in plant tissues (Cichoka, 2015). While indirect damage is related to the fact that aphids serve as vectors in the transmission of viral plant diseases that can kill infected plants or reduce crop yields (Bhoi et al., 2022).

Aphid control has predominantly been carried out using chemical insecticides. However, their intensive use can lead to the selection of resistant populations of the pest, as well as damage to pollinators, natural enemies and animals in the area, not to mention the residues left on food and in the environment, putting human health at risk (Wojciechowska, 2016).

Faced with this situation, in recent years there has been growing scientific interest in the search for and implementation of pest control methods that are less aggressive to the environment, animals and humans (Guerrero, 2013). Some management techniques can be advantageous, especially to avoid future problems such as the development of resistance on the part of the insect. Among the various control strategies are alternative pesticides, which are products prepared from substances that are not harmful to human health or the environment, designed to help control agricultural pests and diseases (Michereff; Lima, 2016).

The possibility of using these recipes in organic production systems is a great advantage and also represents a strategy for the agroecological transition process. This group includes formulations whose main characteristics low or no toxicity to humans and the soil; efficiency in combating arthropods and pests in general; not favouring the occurrence of forms of resistance; good availability and low cost. Included in this category are biocontrol agents, various liquid biofertilisers, sulphocalcic, viçosa and bordeaux syrups, pheromones, plant extracts, among others (Nomura et al. 2018).

In addition to this method, management with entomopathogenic fungi is also effective when used in combination with other insecticides or even alone. When these Microorganisms used on crops are able to regulate insect pests according to their mechanisms of action in a viable, safe and compatible way (Francis et al., 2020). Among the microorganisms used, entomopathogenic fungi such as Metarhizium spp. and Beauveria spp. are widely studied for the development of biopesticides and used to combat various arthropods (OZDEMIR et al., 2020).

Entomopathogenic fungi stand out when compared to other biocontrol systems, such as viruses, protozoa and toxin-producing bacteria, because they have the advantage of having a specialised infection mechanism, which occurs through their active penetration into the hosts, i.e. by contact, thus not depending on their ingestion for the infection process to begin (Fontes, 2020). In addition to the production of secondary metabolites that exclusively cause damage to the pathogen, thus maintaining the population balance in the natural habitat.

This means that natural and biological pesticide technology is viable and safe from an agroecological production perspective. The aim is not only to combat these agricultural pests, but also to maintain the environment and reduce environmental and human contamination. With this in mind, the aim of this study was to evaluate the bioinsecticidal potential of alternative pesticides and entomopathogenic fungi on aphids of the genus Rhopalosiphum, with the aim of contributing to the development of more effective alternatives that are less harmful to the environment.

2.OBJECTIVES

2.1 GENERAL OBJECTIVE

Evaluate the lethal action of alternative pesticides and entomopathogenic fungi on aphids in vitro.

2.2 SPECIFIC OBJECTIVES

- Evaluate the average mortality rate of aphids after infection with M. anisopliae, M. flavoviride, B. bassiana and B. bronginiartii;

- To evaluate the action of Bordeaux mixture, hydrated lime, onion, garlic and red pepper, biofertiliser, manipueira, neem, neutral detergent and cooking oil, paraffin emulsion and soap and coconut soap on aphids;

- Determine the average lethal time (TL_{50}) of the aphids in the presence of the different syrups and entomopathogenic fungi;

- To identify the best control strategy with the potential to be used in integrated pest management in the field.

3.THEORETICAL FOUNDATION

3.1.Aphids and their impact on agriculture

Aphids, also known as aphids or plant lice, are sucking insects that are part of the Aphidoidea superfamily, with the Aphididae family being the most common (Maharani, 2020), with approximately 4,000 species and are found all over the globe. The arrangement of this group represents their great ability to persist under adverse climatic conditions (DAMPC, 2020). They are insects that live in colonies, vary in size from 1.5 to 3 mm in length (Ukhabhai et al., 2022) and their can be different, such as black, red, yellow and green, depending on the species (Figure 1). Although they share an ovoid shape, antennae and legs, some specimens may also have wings.

Figure 1: Phenotypic variation of aphids.

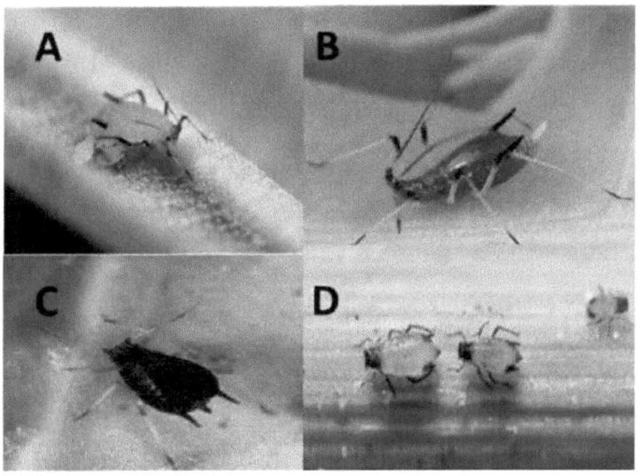

(A): yellow aphid, (B): red aphid, (C) black aphid, (D) green aphid Source: Agrolink, 2019.

The great success of aphids as a pest is due to their high fecundity. Their biological cycle (Figure 2) is quite short, with one female capable of laying up to 50 eggs in a cycle that repeats every three days. In addition to the polymorphism of the individuals that allows them to protect themselves from the combination of temporal and spatial instability, in which, depending on the season or climate, the females can reproduce sexually or asexually, with or without wings, allowing them to adjust their phenotype according to environmental conditions (Simon, 2018; Ziabari, 2023).

Figure 2: Development cycle of Rhopaloshiphum spp.

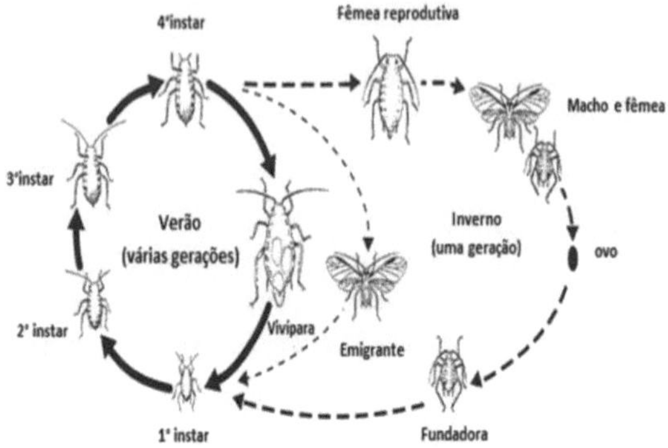

Source: FLINT, 2000.

The majority of reproduction takes place by parthenogenesis, i.e. asexually, in which the females give rise to other females without the need to be fertilised. Development varies according to temperature, where in warmer regions there are several generations a year, by parthenogenesis, while in colder regions reproduction occurs sexually, the length of the life cycle increases and multiplication decreases, giving rise to oviparous males and females (Dampc., 2020). Polymorphism, as proven in the research by Lees (1966), contributes to the adaptation of aphids, as the apterous forms, without wings (Figure 3), concentrate on reproduction in favourable environmental conditions, while the winged forms, with wings, are dedicated to dispersal in adverse conditions.

Figure 3: Aphid apterous (A) and winged (B) forms.

Source: LOPES, 2024.

These insects, both nymphs and adults, feed on sap through suction, causing damage such as the transmission of phytopathogenic agents that reduce production potential, premature leaf fall and favouring the spread of fungi. Other problems resulting from large populations of these insects (Figure 4) include the curving or roughness and distortion of leaves, yellowing, impaired plant growth and dwarfism of shoots (Machado et al., 2020).

Figure 4: Plant infested by aphids (Rhopalosiphum spp.).

Source: Own authorship, 2024.

Aphids have been regularly and intensively exposed to insecticides since the general use of pesticides in the late 1940s. This exposure quickly led to multiple forms of resistance to the main classes of agrochemicals, such as neonicotinoids, organophosphates and pyrethroids, as pointed out in Foster's research (2017). In addition, agrochemicals have a negative impact on the microbiological diversity of the soil and agricultural sustainability, affecting food and nutritional security, generating various damages to the environment and consumers (Meena, 2020).

Agrochemicals impact the soil, i.e. the area that receives most of the complex agrochemical compounds. The vast majority of these compounds are unintentionally poisonous to the activity of beneficial microorganisms, negatively affecting microbial functions and biochemical processes in the soil. As a result, the availability of nutrients is reduced, which contradicts what is trying to be avoided, such as an increase in the incidence of plant diseases and a restriction of plant growth and development due to the poor nutritional status of the soil (Wang et al., 2006; Meena, 2020).

On the other hand, crop fields are monotonous resources with a low diversity of aphid natural enemies compared to their wild habitats (Derocles, 2020). As a result, crops are becoming increasingly infested and there is no truly effective method to combat this pest, which is multiplying extremely quickly.

3.2. The potential of natural pesticides to control pests

In the agricultural context, natural products have been widely applied control rural pests and diseases since the mid-19th century. Preparations based on soap, boron and whale oil were sprayed on plantations (Sparks, 2017). Over time, there was a search for more effective and more powerful products, and so the phase synthetic products began. Over the years, the uncontrolled use of these products has led to problems such as loss of efficiency, a lower spectrum of action and, consequently, food contamination (Tudi, 2021).This has given rise to the need to resume the use of natural, biologically active substances against pests and pathogens, as well as resorting to biological control (JIN et al, 2021). In this context, natural pesticides, also known as "alte matives", are back - products made from substances that are not harmful to the environment.These are measures to control pests and favour the production of healthier food for the end consumer (AYRES, 2020). This class includes plant extracts, oils, natural syrups, pheromones, liquid biofertilisers, entomopathogenic microorganisms, among others.Alternative pesticides have the following characteristics: non-toxicity, low to no aggression towards humans and nature, efficiency in fighting and repelling insects and harmful microorganisms, they do not contribute to possible forms of resistance and are easy to apply and manage (Sharma, 2020).

Replacing synthetic agrochemicals with natural products tends to reduce contamination of water sources, soil, animals and humans, favours plant production and can control pests in a desirable way, such as the study by Halder et al. (2013) where it was found that the combination of neem oil with entomopathogenic fungus had a control of up to 82.47% against ant nymphs of the species Phenacoccus solenopis.Agroecological dressings are being used more and more in agriculture, especially on organic vegetable crops. Their effectiveness has been proven by various studies, such as the work carried out by Spazzin (2023), in which the insecticidal potential of natural mixtures based on garlic, chilli, sage, neem and tobacco against the cabbage caterpillar (Ascia monuste orseis) was assessed. When the effects were assessed, it was observed that the syrups interfered and were generally effective in affecting the emergence of the insects, as well as interfering in the pupal cycle, in which most of the caterpillars did not reach the adult stage, consequently reducing the pest population.Many natural products have desirable properties when it comes to insect control. For example, garlic was used in the Middle Ages as a mosquito repellent when placed on the doors of houses. Its repellent activity is related the presence of secondary metabolites, such as organosulphur compounds, which protect plants against insects and phytopathogens (CARRASCO, 2016; ROUF,

2020). Another case is chilli peppers, which belong to the Piperaceae family, where the plants are a source of unsaturated long-chain isobutylamides with insecticidal properties, such as piperine (JEON, 2019). Its pesticidal activity has been proven in several studies, such as the one carried out by Ashouri (2010) in which the use of black pepper seed powder at a concentration of 0.5% effectively killed adult insects of the species Rhyzopertha dominica and Sitophilus granarius, while red pepper powder reduced their offspring at the highest dosages tested. The benefits of natural dressings are numerous and have a low environmental impact. Most of them are inexpensive, within the farmer's reach and decompose quickly (Fenibo, 2022). For this reason, the technology of these natural dressings is increasingly being used as viable methods within a healthy panorama for living beings in production. For this reason, a solid understanding of the biology, ecology and biodiversity of insects is essential for the implementation of these dressings to be effective, where it is necessary to consider factors such as crop types, agricultural context, knowledge of natural enemies and types of crops to be established (Freires, 2022).

3.3. Entomopathogenic fungi in the control of pests

Entomopathogens include a range of organisms such as protozoa, viruses, bacteria and fungi (Kumar, 2020) that can cause a series of disorders and diseases in insects. The way they are infected varies depending on the organism, with the possibility of it being through the digestive tract or cuticular routes (Mannino, 2019).In particular, entomopathogenic fungi naturally play a fundamental role in regulating insect populations in nature. This is why they were the first biological agents to be used to control economically important insects. In addition, they have a broad spectrum of action, being able to colonise various species: flies, grasshoppers, aphids, caterpillars, beetles, etc (Wang, 2017; Islam, 2021). The genera Metarhizium, Beauvaria, Entomophthora, Aspergillus and Paecilomyces stand out as the main entomopathogenic fungi, acting as a mechanism to regulate pest populations in the biosystem (Bezerra; Sousa, 2018). Among these, the ones that stand out the most are Beauvaria spp. and Metarhizium spp. for presenting a range of studies and results demonstrating their potential as microbial control agents, being widely used in various crops (Sofwah, 2023).

3.3.1 Beauvaria spp.

Beauvaria spp. is considered an anamorphic cosmopolitan genus of arthropod, pathogenic and necrotrophic soil-borne fungi (Bustamante, 2019). Morphologically, Beauvaria spp has been characterised asexually by having conidiogenous cells arising from short peduncles, often unicellular, with more or less swollen cells, and most of the time in dense clusters, or mirrored or in spirals of undifferentiated hyphae (De Hoog, 1972). In addition to hyaline, smooth, thin-walled globose to ellipsoidal unicellular conidia. The sexual forms form solitary, paired, unbranched stromata, fleshy texture, apical fertile area, cylindrical to clavate (Kepler et al., 2017).Among the Beauvaria species, B. bassiana and B. brongniartii are the main ones in the agronomic context of biological combat, widely used as mycoinsecticides. Morphologically, there are differences in the shape and size of the conidia (Figure 5), where B. bassiana has mainly spherical, hyaline and smooth conidia, simple and irregular conidiophores and globose or bulky phialides, which are thinner at the top and the colonies are flattened, mealy and finely pulvurulent. While B. brongniartii, the conidia are broader and cylindrical, ellipsoid, conidiophores scarce and rarely in clusters, phialides when compared to B. bassiana are less broad, with a thinner base with a ventriculose or filamentous shape and their colonies have a whitish to creamy yellow appearance (Jaronski, 2017; Castillo et al, 2013; Fan et al, 2013).

Figure 5: Conidiophores and conidia of Beauvaria bassiana (A); conidiophores and conidia of Beauvaria brongniartii (B).

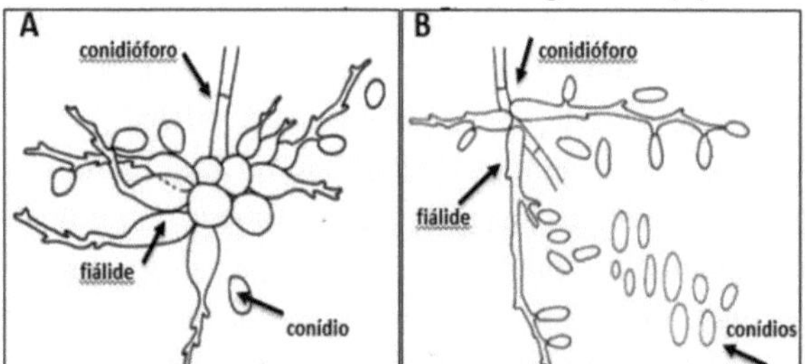

Source: CASTILLO et al. 2013.

In addition, these two species have different specificities, while B. bassiana has a very diverse range and can infect an extensive range of hosts, including beetles, flies, locusts, among others (Devi, 2008), B.brongniartii tends to be more specific, being particularly effective against beetle larvae, especially from the Scarabaeidae family. An example of research showing its applicability against this family is the study by Nong (2011) in which isolates of B. brongniartii caused around 80 per cent mortality in Anoma corpulenta larvae, suggesting its potential in preventing larval infestations in various crops. Despite their differences, both have the same general basic mechanism of infection, which are divided into: adhesion and germination, formation of appressoria, formation of penetration structures, colonisation and reproduction of the fungus (Figure 6) (García et al., 2008).

Figure 6: Mechanism infection and pathogenicity of entomopathogenic fungi.

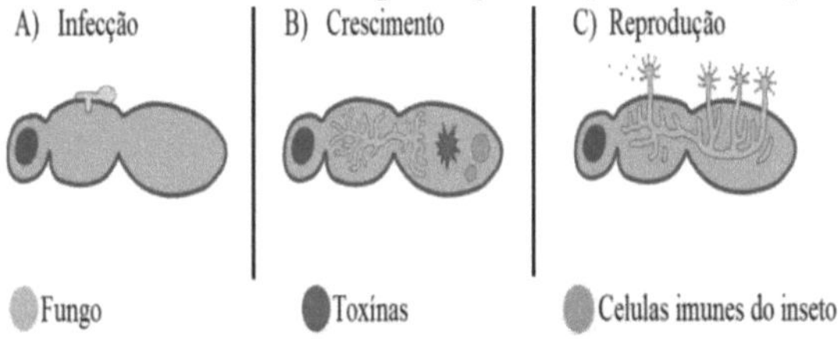

Source: Souza, F. J., 2021.

The process begins when the spore or conidium adheres to the insect's cuticle, after which a germ tube and an appendage are produced. The appressoria are swellings in the apical region of the germ tubes, which are responsible for anchoring the conidia, thus attaching them to the cuticle with the germ tube so that penetration into the insect's body can then take place, both physical and chemical mechanisms take place. The first consists of the pressure exerted by the hyphae, which breaks the membranous areas of the cuticle (Ortiz, 2013).The chemical mechanism consists of enzymatic action, mainly proteases, lipases and chitinases, which cause tissue decomposition in the penetration zone. After penetration, the hyphae enlarge and branch out within the insect's tissue, completely colonising the tissue. from which small colonies and structures of the fungus form, which corresponds to the final stage of the insect's illness (Tennakoon, 2021). Affecting the host insect involves both direct contamination by the fungus and the action of the secondary metabolites it produces, some of which are toxic to the pest (Bezerra, 2018). Toxic metabolites include extracellular enzymes, proteins and low molecular weight compounds. In the case of the genus Beauveria spp, the mycotoxins bassianolide and beauvericin produced by B. bassiana stand out. The second toxin has been highly studied and, currently, research such as that by Khoury et al (2019) has shown that the substance present in the fungus causes 100 per cent mortality when used in concentrations of 100 and 1000µg/g in varieties of mites. In this way, infection with the fungus causes insects to suffer a significant drop in their immune systems. Some indications of illness are dark spots, lack of coordination, paralysis, altered behaviour and reduced thermal tolerance, all of which change the tendency of insect species. The insect becomes debilitated when it stops feeding and its death occurs a few days after contamination with the fungus thanks to a series of factors ranging from nutrient depletion to physical damage caused by mycelial vegetative growth (Fan et al., 2013; Porras, 2021).

After the host dies, the fungus breaks through the cuticle once again and sporulates on the surface of the corpse. Insects colonised by Beauveria spp start to show a specific white colour (Figure 7) characteristic of the fungal species, since after death they become mummified by the mycelial mass of the fungus (PASTOR, 2013).

Figure 7: External sporulation of Beauveria bassiana.

Source: Oklahoma State University, 2021.

This mycelial mass has conidiophores which, through their conidiogenic regions, produce new conidia that can be spread either by individuals who have had the disease or by wind or rain, allowing the biological cycle of the fungus to continue (Lage et al., 2013).

The effectiveness of Beauveria spp. as a biological pest control agent is the reason why this fungus has been widely used in various agricultural crops such as soya, maize, coffee, among others. In Brazil, it is mainly used to control leafhoppers in sugarcane fields and whiteflies, but its market share remains small compared to chemical insecticides. In China, thousands of tonnes of fungal biocontrol formulations are applied every year to control insect pests, and it is estimated that around one million hectares per year are treated with Beauveria by the Chinese government (Xiao et al., 2012; Wang, 2014). This reveals the high potential already exploited by these fungi in the agricultural field.

3.3.2Metarhizium spp.

Of great global renown, the genus Metarhizium stands out as a notorious pest control agent and has been successful in controlling arthropods in Brazil, acting naturally on more than 300 species (Mesquita, 2023). This genus of entomopathogenic fungi has been widely studied due to its specificity to various hosts, ease of mass production and environmental safety (Rumbos, 2017).

The genus Metarhizium is distributed globally from the Arctic to the tropics, but is most abundant in pastures followed by forests (Fernández-Bravo, 2021). This genus belongs to the class of Hyphomycetes, which reproduces by spores known as conidia and has been isolated from infected insects and soil on every continent except Antarctica.

The most important taxonomic revisions on Metarhizium were made by Margaret Tulloch in 1976, where two species were distinguished: M. anisopliae and M. flavoviride. Currently, M. anisopliae is the most widely produced and used entomopathogenic fungus in Brazil, with several isolates suitable for biological control and being pathogenic to dozens of species of insects from different orders (Mascarin, 2018).

The name of the species comes from the fact that it was first isolated by the Russian Ilya Metchnikoff from the beetle Anisoplia austriaca, and the fungus was given the name Entomophthora anisopliae. In 1883, Pitirim Soroking classified it in the genus Metarhizium where it remains to this day. The first work using the fungus as a biocontrol was in 1879 when the fungus was applied to the larvae of a curculionid, a beetroot pest (Alves, 1998).

Morphologically, the conidiophores of M. anisopliae are simple, in which each conidiophore forms a chain of conidia that grow densely and adhere to each other forming prismatic basipetalous masses. The conidia are cylindrical to oval in structure, varying in length according to the variety, from 3.5 to 18 μm long. The phialides are cylindrical or clavate. The fungus has hyaline, septate mycelium and veined collar spores, causing what is known as "musca dime ve de" on insects (Figure 5) (Bezerra, 2018; Jeffs et al., 1999).

Figure 8: Bedbug species infected by Metarhizium anisopliae.

Source: Livestock Circuit, 2021.

According to Alves (1998), they can be differentiated based on the size of the conidia (Figure 5), where short conidia measuring 3 to 9 µm are of the minor variety, while long conidia measuring 9.1 to 18 µm are of the majus variety (Figure 9). The colour of the conidiophores of the M. anisopliae species can be white, greenish, brown or light brown (Lopes, 2016)

Figure 9: Structure of Metarhizium anisopliae conidia - (A) Metarhizium anisopliae var. minus; (B) Metarhizium anisopliae var. majus.

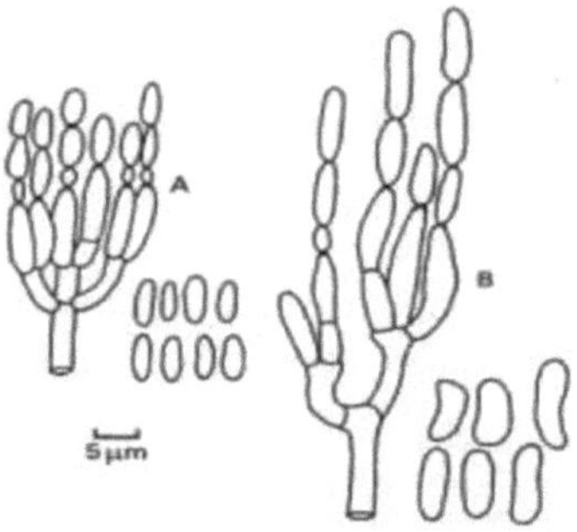

Source: ALVES, 1998.

With regard to the species M. flavoviride, Gams & Rozsypal (1973) used the morphology of the conidia and conidiophores to describe this new fungus, which was isolated from curculionids in 1956 from cultivated soils in several European countries. M. flavoviride is characterised by the formation of conidiophores originating clavate phialides and chains of ellipsoid to cylindrical conidia depending on the time of the strain (young to old) as recognised by Driver et al. (2000) (Figure 10).

Figure 10: Structure of Metarhizium flavoviride conidia.

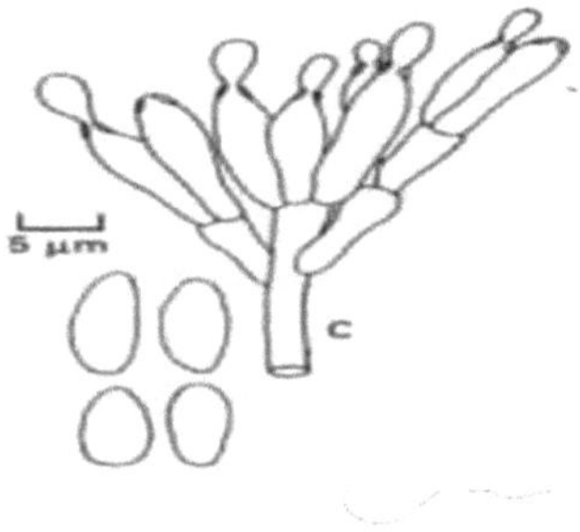

Source: ALVES, 1998.

Regarding the morphology of the fungus, the colony is coloured white to green and has a velvety texture. Its conidiophores are erect, simple or branched, emerging directly from the mycelium and varying in colour from green to olive green. The conidia are hyaline, cylindrical to elliptical in shape and are produced in chains at the tips of the conidiophores (Hodges, 1997).

M. flavoviride has been used extensively for biological control, mainly of Orthoptera, Homoptera and Coleoptera, where it has been tested on various insects and has shown satisfactory results, such as the research published by Vivekanandhan et al. (2022), in which the virulence of M. flavoviride conidia against Lepidoptera was studied. It was found that the conidia cause harmful effects on the haemocyte levels of the larvae, as well as modifications to the antioxidant and detoxifying enzyme profiles, generating mortality of up to 80% for the S. litura species.

Although studies have been carried out on the factors that can influence the sporulation of this fungus (Magalhães; Frazão, 1996), currently little is known about the species and its parasexual cycle, which has not yet been described, making it impossible to study its genetics and consequently the mechanisms of pathogenicity and recombination (Lima 2005). Despite being different species, both fungi described: M. anisopliae and M. flavoviride present the same infection process that occurs in successive phases of germination, differentiation, penetration, colonisation, reproduction and dissemination (Mannino, 2019). The mechanism begins with the germination of the spores on the host's cuticle and the formation of appressoria (turgor pressure cells) that break through the cuticle (Pedrini, 2018).Once inside the haemocele, the fungal cells acquire yeast-like forms, called hyphal bodies, which invade the host through a sequential process, ending in its mummification. Metarhizium fungal cells can also secrete toxic secondary metabolites, such as the toxin destruxin, which has immunosuppressive and paralysing effects, facilitating fungal infection (Barelli, 2022).In terms of the effectiveness and application of both species, M. flavoviride is particularly effective against grasshoppers and other orthopterans. It is often used in tropical and subtropical regions. M. anisopliae, on the other hand, is widely used in various parts of the world against various insects and is applied in different agricultural and forestry contexts. Although they share a similar genetic heritage, the genetic differences and Their specific biochemical characteristics allow them to adapt to different hosts and environments, and both are suitable for use in biological control programmes (Iwanick, 2020; TIAGO, 2002).

4.MATERIAL AND METHODS

The experiments were conducted at the Laboratory of Molecular Genetics and Plant Biotechnology (LAGEMOL), Biotechnology Centre, Federal University of Paraíba (UFPB), Campus I, João Pessoa/PB and Empresa Paraibana de Pesquisa, Extensão Rural e Regularização Fundiária - EMPAER, located at Rodovia Abelardo Jurema, PB 008, Km 7, Jacarapé III, s/n, João Pessoa/PB.

4.1. Origin of the insect: the aphids were provided by the Paraibana Rural Research and Extension Company (EMPAER) (Figure 11).

Figure 11: Aphids used in the different tests.

Source: Own authorship,

4.2. The origins of the fungal strains: Metarhizium anisopliae (Access URM 6133), Metarhizium flavoviride (Access URM 3998), Beauveria bassiana (Access URM 2915) and Beauveria brongniartii (Access URM 6504) were provided by the URM Mycoteca of the Department of Mycology/Federal University of Pernambuco/UFPE (Figure 12).

Figure 12: Colonies of Metarhizium anisopliae (A), Metarhizium flavoviride (B), Beauveria bassiana (C) and Beauveria brongniartii (D) on 2 % Sabouraud-dextrose agar culture medium.

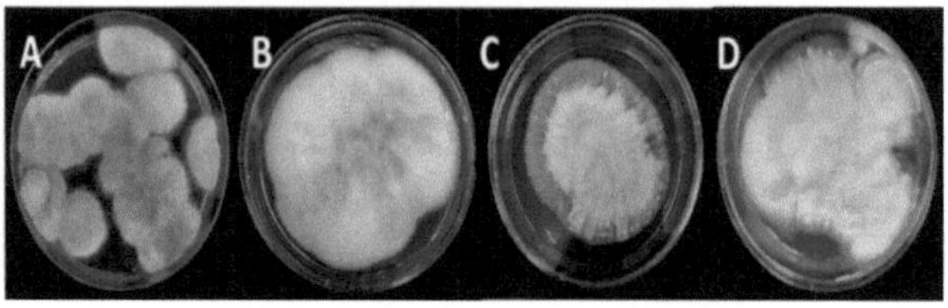

Source: Own authorship, 2024.

4.3.Culture medium for maintenance and conidiogenesis of the fungal strains: 2% Sabouraud-dextrose agar (5 g/L meat peptone, 20 g/L glucose, 5 g/L casein peptone and 15 g/L bacteriological agar. pH 5.6 ± 0.1 at 25 °C). The samples were transferred to test tubes and Petri dishes containing 2% Sabouraud-dextrose agar medium, where the cultures were kept at room temperature for 10 to 15 days and then refrigerated at 4 °C.

4.4. Types of natural syrups: Bordeaux, hydrated lime, onion, garlic and red pepper, manipueira, neem, neutral detergent syrup and cooking oil, paraffin emulsion, biofertiliser and soap and coconut soap.

4.5.Preparing syrups

4.5.1. Bordeaux mixture: 1 litre of clean running water, 5 grams of copper sulphate and 8 grams of hydrated lime. The volume of water was divided into two 500 mL parts in a beaker. 5 g of copper sulphate was poured into Beaker 1 and 8 g of hydrated lime into Beaker 2. two beakers were then shaken and then Beaker 1 was slowly poured into Beaker 2 and mixed with a glass rod (Figure 13). The pH was checked and adjusted to 6.5 or 7.0. Once ready, the mixture was applied to the pest using a spray bottle (Penteado, 1996).

Figure 13: Bordeaux mixture.

Source: Own authorship, 2024.

4.5.2. Hydrated lime slurry**:** 1 kg of quicklime and 8 litres of water. The lime was weighed and then half the volume of running water was added. A slightly soft paste was formed when the lime was heated and an exothermic reaction took place. After cooling the powder, the rest of the water was added to obtain a milk of lime (Figure 14). The mixture was left to stand for 24 hours and then applied to the pest using a spray bottle (Lima, 1993).

Figure 14: Hydrated lime grout.

Source: Own authorship, 2024.

4.5.3. Onion, garlic and red pepper syrup: 100 grams of red pepper, 1 litre of water, 200 grams of onion and 4 grams of garlic clove. The chilli, onion and garlic were ground in a blender in 500 mL of water. It was then strained to remove the juice and mixed with the rest of the water (Figure 15) and then applied to the pest using a spray bottle (Elwell; Maas, 1995).

Figure 15: Onion, garlic and red chilli syrup.

Source: Own authorship, 2024.

4.5.4. Manipueira syrup: (milky juice extracted when the grated cassava is squeezed): 1 kg of grated cassava and 2 litres of water. The cassava was left to soak in the water for 24 hours and then strained through a cloth to recover the milky juice (Figure 16). Once ready, it was applied with a spray bottle within 24 hours (Elwell; Maas, 1995).

Figure 16: Manipueira syrup.

Source: Own authorship, 2024.

4.5.5. Neem syrup: 5 mL **of neem** oil was diluted in 1 L of running water, homogenised and sprayed (Figure 17) (Abreu, 1998).

Figure 17: Neem syrup.

Source: , 2024

4.5.6. Neutral detergent and cooking oil: 500 mL of neutral detergent and 500 mL of cooking oil. The two ingredients were mixed and shaken well in a container. 1% was applied to young plants and 2 to 3% to adult plants (200 to 300 mL/10 L of running water) (Figure 18). The mixture was applied to the pest using a spray bottle (Guerra, 1985).

Figure 18: Neutral detergent syrup and cooking oil.

Source: Own authorship, 2024.

4.5.7. Paraffin and soap emulsion: 1 litre of paraffin, 1 litre of water and 400 grams of ordinary soap. 1 litre of water was mixed with the 400 grams of soap and the mixture was heated until the soap melted. When it boiled, it was removed from the heat and mixed with 1 litre of paraffin, stirring constantly, until the paraffin was completely mixed with the water and soap, forming a light grey creamy paste. It was stirred until it cooled (Figure 19). The mixture was then applied to the pest using a spray bottle (Guerra, 1985).

Figure 19: Kerosene and soap emulsion syrup.

Source: Own authorship, 2024.

4.5.8. Coconut soap syrup: 5 grams of grated coconut soap and 1 litre of running water. The grated coconut soap was diluted in approximately 100 mL of warm water. The remaining 900 mL of clean water was then added (Figure 20). The mixture was put in a spray bottle and applied to the pest (Guerra, 1985).

Figure 20: Coconut soap syrup.

Source: Own authorship, 2024.

4.5.9.Biofertiliser syrup - milk, ash, fresh cow dung **and sugar:** 40 L of cow dung, 100 L of water, 1 L of milk and 1 L of molasses were added to a 200 L container (Basic Ingredients 1). It was mixed well and left to ferment for three days. Every five days, one of the mineral salts (Mineral Salts 2) was dissolved in 2 litres of warm water and added to 1 litre of milk; 1 litre of molasses (or 0.5 kg of sugar) and one of the complementary ingredients (Alternative Complements 3) and mixed with the fermenting manure. Then all the mineral salts (Mineral Salts 2) were added in the suggested order, making up to 180 litres. The container was sealed to ferment for thirty days in summer or forty-five days in winter, before the mixture was used. After this period, the mixture was sprayed on the leaf area at a concentration of 1 to 5% (Figure 21) (Três; Resende, 1995). Basic ingredients 1: Fresh cow dung 40 litres; water 140 litres, milk/whey 9 litres and molasses 9 litres. Mineral salts 2: Zinc sulphate - 3 kg, Magnesium sulphate - 1 kg, Manganese sulphate - 0.3 kg, Copper sulphate - 0.3 kg, Calcium chloride - 2 kg, Borax (or boric acid 1.0 kg) - 1 kg, Cofermol (Cobalt, Iron and Molybdenum) - 0.125 kg.Alternative supplements 3: Bone meal - 0.2 kg, fish scraps - 0.5 kg, blood - 0.1 kg and ground liver scraps - 0.2 kg.

Figure 21: Biofertiliser syrup.

Source: Own authorship, 2024.

4.6. Bioassay with natural syrups: the methodology used for the pathogenicity test was adapted from the work of Santos et al. (2007). An experiment was carried out with 40 specimens of insect pests, corresponding to three replicates of 10 insects for each treatment (TC Control group = without the mixture; Treatments 1, Treatments 2 and Treatments = The insects were sprayed with the different mixtures). The insects were transferred in groups of 10 per glass jar containing filter paper at the bottom to absorb moisture and the diet (leaves from the crops from which the insects were collected (Figure 22). The syrups were applied to the insects every five days. After this, the glass jars were kept at a temperature of 25 ± 2 °C and evaluated every 24 hours for 20 days to confirm the death of the insects in the presence of the different natural syrups.

Figure 22: Structure of the in vitro bioassay with aphids.

TC: Control treatment - 10 aphids without syrup. Only distilled water was sprayed; T1, T2 and T3: each treatment contained 10 insects and was sprayed with syrup. Source: Authors, 2024.

4.7. Bioassay with entomopathogenic fungi

4.7.1. Preparation of conidial suspensions: the fungi were inoculated onto 2% Sabouraud-Dextrose Agar culture medium. The plates were kept at a temperature of 25 ± 2 °C for a period of 10 to 15 days for growth and conidiogenesis. After this period, the conidia were collected by scraping the surface of the colony to carry out the bioassays. Autoclaved distilled water and 0.01 % (v/v) Tween 80 were used to prepare the conidia suspensions. The concentration of conidia was then estimated in a Neubauer chamber and the suspensions ranged from 1.0 x 10^5 conidia mL^{-1} to 1.0 x 10^6 conidia mL^{-1}.

4.7.2. Pathogenicity test: the methodology used was adapted from the work of Santos et al. (2007). An experiment was carried out with 40 aphid specimens for each fungus species, corresponding to four replicates of 10 insects for each treatment (TC control group = autoclaved distilled water + Tween 80; T1 = B. bassiana conidia suspension + Tween 80; T2 = B. brongniartii conidia suspension + Tween 80; T3= Mnsion+ Tween 80 an=suspension Tween 80 and T4 M. aniTween 80). M. flavoviride + Tween 80) (Figure 23). The insects were transferred individually to Petri dishes containing filter paper with 1 mL of the conidia suspension and the diet (coconut palm leaves). The plates were kept at a temperature of 25 ± 2 °C and evaluated every 24 hours for 10 days to observe the extrusion of the fungus and confirm death by the pathogen.

Figure 23: Bioassay to assess the pathogenicity of the fungal strains.

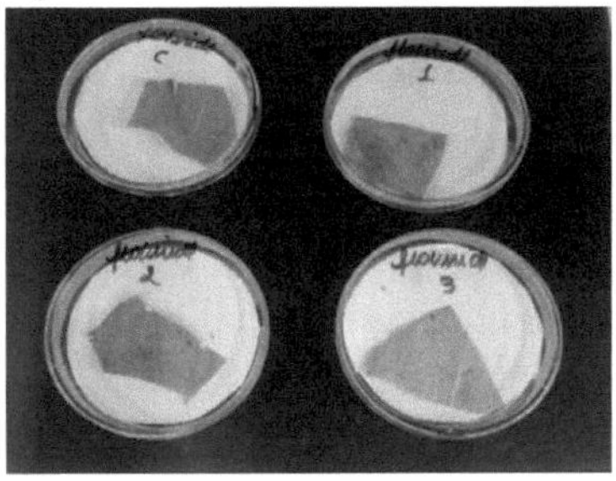

Source: , 2024

4.8. Statistical analyses: the experiments were carried out using a completely randomised experimental design, where the data was statistically analysed for variance (F test) and the means were compared with each other using the Tukey test at a 5% probability level, using the Sisvar software (Ferreira, 2014). The data relating to confirmed mortality was subjected to Probit analysis to obtain TL_{50} (lethal temperature) values in days (Finney, 1971).

5.RESULTS AND DISCUSSION

5.1.Characterisation of the conidia of the fungal strains

Microscopic analysis of the four fungal species showed that all of them had preserved and viable structures. With regard to M. anisopliae (Figure 24A), the cylindrical, elongated and uninucleate morphology was preserved as described by Jeffs et al. (1999), with dimensions varying from 3.5 µm to 17.7 µm (Riba et al., 1986). M. flavoviride (Figure 24B), on the other hand, differs in the shape of its conidia, with the sexual spores having a flatter elliptical/cylindrical shape and a smaller size when compared to M. anisopliae (Driver, 2000). In relation to the strains of the genus Beauveria (Figures 24C and 24D), both showed asexual spores that were quite small in size, measuring between 2 µm and 4 µm in length and with a round globose shape, which are arranged on the stems of the phialides. In the case of B. bassiana, the conidia are produced in conidiophores that form clusters in the shape of racemes. In the case of B. brongniartii, the conidiophores are sparse, rarely in clusters and the conidia appear more dispersed, but still grouped and close together, confirming the characteristics reported by Humber (1997) and Zimmermann (2007).

Figure 24: Conidia of Metarhizhium anisopliae (A), Metarhizium flavoviride (B). Beauveria bassiana (C), Beauveria brongniartii (D).

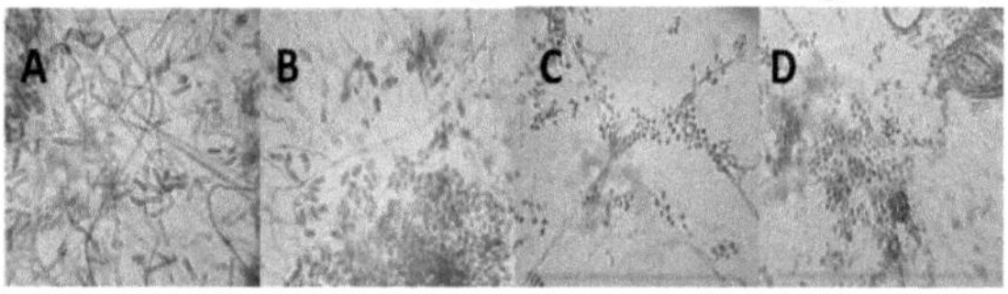

Source: Own authorship: 2024.

The characterisation and analysis of the conidia of fungal strains is extremely important in order to determine the integrity, viability and vigour of the structures. are the propagules responsible for starting the infection process on the insect's cuticle (Jeffs et al., 1999). Therefore, a low quantity of conidia can reduce the efficiency of the fungus in relation to the host. It is therefore necessary to ensure that the control agents are viable and capable of infecting the target organism (Alves, 1998).

5.2. Evaluation of the pathogenicity of fungal strains on aphids

The pathogenicity of B. bassiana (1x10^5 conidia mL^{-1}), B. brongniartii (1x10^6 conidia $mL^{(-1)}$)), M. anisopliae (1x10^6 $mL^{(-1)}$)) and M. flavoviride (1x10^5 conidia $mL^{(-1)}$)) on aphids was analysed (Table 1). The parameters considered for the pathogenicity of the fungi were the total mortality rate and the mortality rate confirmed by the fungus, since it is only possible to confirm the infection of the pathogen through the presence of external sporulation or conidial mass of the infected insects.

Table 1. Efficacy of Beauveria bassiana (10^5 conidia mL^{-1}), Beauveria brongniartii (10^6 conidia $mL^{(-1)}$), Metarhizium anisopliae (10^6 conidia $mL^{(-1)}$)) and Metarhizium flavoviride (10^5 conidia $mL^{(-1)}$)). Incubation period - 15 days (T = 28 ± 2 °C) (n= 30 for each fungus species plus the control group).

Treatments with fungi	Total mortality rate (%) (±EP)(1)	Mortality rate confirmed by the fungus (%) (±EP)(1)
TC - Control	0.00± 0.00 b	0.00± 0.00 b
T1 - Beauveria bassiana	96.4± 12.34 a	96.4± 6.49 a
T2 - Beauveria brongniartii	100.0± 10.99 a	100.0± 9.79 a
T3 - Metarhizium anisopliae	93.9± 7.08 a	00.0± 0.00 b
T4 - Metarhizium flavoviride	99.0± 4.29 a	00.0± 0.00 b
CV%	15,29	5,65

1Standard error: Averages with the same letter do not differ according to the Turkey test at 5%. Source: Own authorship, 2024.

The total mortality rate in all treatments above 90%, with B. bassiana 96.4%, B. brongniartii (100%), M. anisopliae 93.9 and M. flavoviride 99%. However, with regard to the rate of confirmed mortality by the fungus, only the treatments with the two species of Beauveria showed colonisation and extrusion of the mycelium on the insect with a rate above 96% of confirmed mortality for B. bassiana and 100% for B. brongniartii. For the two Metarhizium species tested, there were no signs of fungal colonisation. The pathogenicity and virulence of fungi on insects is proven when the mycelium colonises and externalises with a conidial mass, leading to mummification of the insects. The lack of colonisation of M. anisopliae and M. flavoviride on the aphid can be explained by the high natural variability of this genus. Despite its high applicability against various pests such as termites, grasshoppers, leafhoppers and beetles (including the larval stage), the genus Metarhizium has been studied from a physiological and molecular point of view to assess its entomopathogenic potential (Humber,

2003). The cycle of fungus-host relationships depends on environmental conditions (temperature, humidity, light and ultraviolet radiation), nutritional conditions and host susceptibility. Most entomopathogenic fungi require at least 95 per cent relative humidity on the insect's surface for germination, germ tube extension and infection (Chuquibala-Checan et al., 2023). The absence of at least one of these factors may have interrupted the infection process of Metarhizium spp. in aphids or even the lack of specificity.

Entomopathogenic fungi produce hydrolytic degrading enzymes that help penetrate the insect cuticle and use the more substantial chitin layer as a substrate. As aphids have a thinner cuticle and less chitin, their effectiveness is reduced. Therefore, the pathogenicity of M. anisopliae and M. flavoviride is highly associated with the secretion of enzymes that degrade the host's cuticle (Santi, 2010). This other factor may also have influenced the absence of Metarhizium spp colonisation on the aphids. Beauveria spp. produce toxins such as beauvericin and bassiana that help parasitise and kill their hosts. Their mechanism of pathogenicity varies according to the toxin released and the type of insect. However, the consensus is that the insecticidal effect is a cumulative result of several toxins, especially in the case of B. bassiana. Insecticidal mechanisms range from the proliferation of virulence factors, impeding the activation of the immune system and interrupting nerve conduction pathways, damaging the epidermis of the host to help the hyphae penetrate and clog the spiracles (Ortiz-Urquiza, 2013; Pedrini et al., 2013; Wang, 2021). In the present study, B. bassiana and B. brongniartii showed virulence and pathogenicity on aphids. Figure 25 shows the complete mummification of the aphid with B. bassiana and the initial colonisation by B. brongniartii. Similar results were reported by Bezerra et al. (2018) using the entomopathogenic fungi Beauveria bassiana, Beauveria brongniartii, Metarhizium anisopliae, Paecilomyces sp. and Aspergillus sp. to control leaf-cutting ants (Hymenoptera: Formicidea).

Figure 25: Aphids colonised by Beauvaria bassiana (A); Beauvaria brongniartii (B).

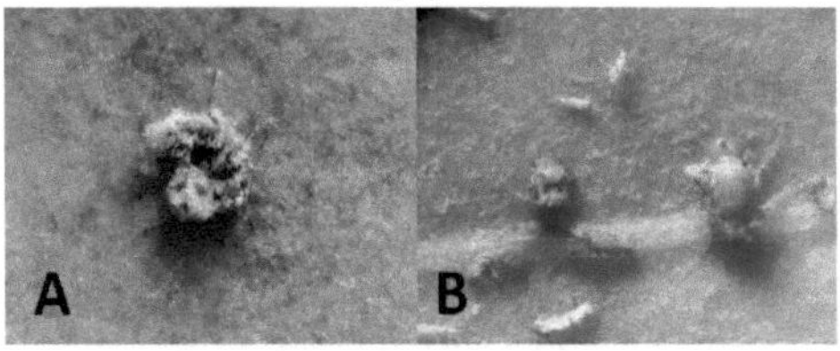

Source: Own authorship, 2024.

5.3. Lethal time (TL_{50}) of fungal strains on aphids

TL_{50} refers to the period required to cause the death of 50 per cent of infected hosts. Evaluating this parameter makes it possible to ascertain the capacity and virulence of these strains (biancalani, 2018).The lethal time for B. bassiana and B. brongniartii was five and six days, respectively (Table 2). There was no lethal time for the insects tested with the fungi of the Metarhizium species because they did not show mortality confirmed by the colonisation of the insect fungi, so there was a significant TL_{50}. value.

Table 2: Lethal time (TL_{50}) for aphids infected with Beauveria bassiana ($1x10^5$ conidia mL^{-1}), Beauveria brongniartii ($1x10^6$ conidia mL^{-1}), Metarhizium anisopliae ($1x10^6$ conidia $mL^{(-1)}$)) and Metarhizium flavoviride ($1x10^5$ conidia $mL^{(-1)}$)). Incubation period - 15 days (T = 28 ± 2 °C) (n=30).

Treatments with fungi	TL50 days) (IC 0,05)4	Regression equation1	t2
TC - Control	0	ns3	0
T1 - Beauveria bassiana	5 (5,02-7,34)	Y=0.1529+ 1.1330 log X	4,72
T2 - Beauveria brongniartii	6 (2,81-4,31)	Y=0.4698+ 1.6512 log X	5,49
T3 - Metarhizium anisopliae	0	ns3	0
T4 - Metarhizium flavoviride	0	ns3	0

[1]Values of the equations obtained by probit analysis, where Y = probit and X = concentration. [2]The value of t determines the angular coefficient; values less than 1.29 indicate that the regression is not significant. [3]Not significant. [4]Confidence interval significant at 5%. CV, coefficient of variation. Source: Own authorship, 2024.

The variation of days in the lethal time of a fungus will depend on a series of factors such as the species of insect used and its stage of development, the type of fungus, its variety and the concentration of conidia applied, as well as environmental conditions such as humidity and temperature. The variation in virulence between isolates, species and doses is determined more by quantitative than qualitative differences in fungal growth kinetics (Anderson, 2011; Férnandez-Conradi, 2018). For this reason, their average time values vary widely in the literature and are used to correlate with the mortality rate. In the case of the study in question, the values for mortality and TL_{50} had a positive correlation, in that as the lethal time increased, there was also an increase in the insect's mortality rate.

5.4. Lethal action of syrups on aphids

The syrups were applied every five days for a period of 20 days to check the average mortality rate and average lethal time (Table 3), during which time the aphids were monitored and kept under controlled laboratory conditions and assessed for any signs of lethargy and probable fading.

Table 3: Mean mortality (± SE) and mean lethal time (TL_{50}) for bioassay with natural syrups for aphid control under laboratory conditions. Evaluation period - 20 days (Ambient temperature= 25 ± 2 °C) (n=40 for each insect).

Caldas	Rate of mortality (%)	TL50 (days) (CI 0.05)4	Equation of regression1	t2
	(±EP)1			
Bordeaux	100.0± 13.60	3 (1,89-2,68)	Y= 0.1782+ 2.8522 log X	2,85
Hydrated lime	100.0± 10.02	3 (2,42-3,04)	Y=2.9145+ 1.3001 log X	5,01
Onion, garlic and chillired	100.0± 9.93	1 (4,91-6,75)	Y=2.1907+ 1.1321 log X	3,09
Manipueira	80.0± 5.78	1 (9,34-11,45)	Y=1.6001+ 5.1725 log X	2,83
Nim	100.0± 3.50	1 (3,32-5,01)	Y=1.2374+ 2.2291 log X	6,75
Neutral detergent and kitchen	100.0± 4.98	1 (6,9 -8,6)	Y=2.2004+ 3.1041 log X	4,80
Soap and paraffin emulsion	100.0± 11.90	1 (10,29-13,75)	Y=0.9401+ 4.1945 log X	3,67
Coconut soap	100.0± 07.21	3 (7,56-10,98)	Y=1.5894+ 2.3381 log X	3,05
Biofertiliser syrup	99.0± 2.69	3 (3,79-6,33)	Y=3.0220+ 0.8034 log X	2,64
Control	20.0± 0.10	0	ns3	0
CV%	10,1			

[1]Values of the equations obtained by probit analysis, where Y = probit and X = concentration. [2]The value of t is crucial for determining the angular coefficient; values below 1.96 indicate that the regression is not significant. [3]Not significant. [4]Confidence interval significant at 5%. CV, coefficient of variation (%). Source: Own authorship, 2024.

In relation to the bioassays carried out, the average mortality rate did not differ between the treatments, in which the values ranged from 80-100%: 80% (manipueira), 99% (biofertiliser syrup) and 100% mortality (Bordeaux mixture; hydrated lime; onion, garlic and pepper; neem; detergent; soap and paraffin emulsion; and coconut soap). Eight of the nine dressings tested showed results above 98 per cent, proving to be highly effective controlling this insect. Thus, the syrups with the most promising results were those that showed 100% mortality in the shortest average lethal time, in this case the syrups: onion, garlic and chilli; neem; detergent and soap and paraffin emulsion.

Due to the variety of compounds, each mixture has a particular characteristic that has proved effective in combating these insects. In the case of onion, garlic and chilli syrups, the compounds have properties that are responsible for suffocating and intoxicating the insect, such as the capsaicin present in chilli and the organosulphur compounds in garlic (Souza, 2020; Oliveira, 2017). When applied to the surface of insects, they break down the lipid layer that protects the pest's skin, causing it to lose water and eventually dehydrate and die (Cavalcanti et al., 2021; Moura, 2018). In general, the aphids showed a high sensitivity to the natural syrups (Figure 26), since the average lethal time of the syrups varied between 1-3 days. These times are relatively low when compared to the study by Menezes (2019) where algarobeira extract was used as an alternative to control black aphids and mortality was 5.94% in four days.

In turn, the use of alternative insecticides such as mastic extract to control aphids carried out by Bezerra (2023) showed a mortality curve of up to 62.85% of adult insects after five days. Not to mention that formulations of aqueous extract of chilli fruit and neem leaves on the mortality of B. brassicae in the evaluation carried out by Botti (2015) showed a maximum mortality of 35.25% on 1% concentration of chilli fruit extract. This means that the values obtained in this study were promising and more efficient than those currently found in the literature for combating aphids, proving to be viable alternatives for controlling these insects.

Figure 26: Macroscopic appearance of the dead aphid, after spraying with the different natural syrups.

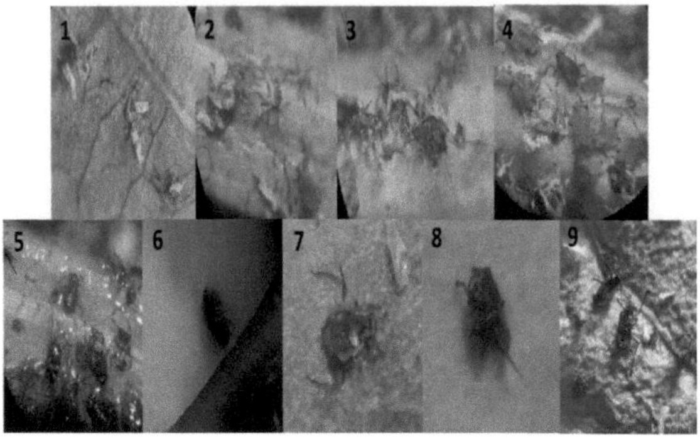

1-Hydrated lime; 2- Bordeaux mixture; 3- Coconut soap mixture; 4- Neem mixture; 5- Detergent and oil mixture; 6- Kerosene and neutral soap mixture; 7- Onion, garlic and pepper mixture; 8- Manipueira mixture; 9- Biofertiliser mixture. Source: Own authorship, 2024.

Currently, various natural products are being combined and used as viable alternatives against pests that attack agricultural production, such as the work by Holtz (2023) in which the efficiency of ripe lemon peel was tested in the control of the spider mite (Raoiella indica) with a mortality rate of 88 per cent at a concentration of 12 per cent of the extract, thus offering a more sustainable and safer alternative for the control of this pest. Thus, despite the fact that syrups are not yet associated with significant and immediate effects for controlling insect pests, recent years of research in the area of control with alternative insecticides have shown great changes through the use of natural products to replace pesticides (Parra; Coelho, 2019). In addition to providing effective solutions, they also offer a more sustainable and safer option, contributing to protecting the environment, human health and maintaining food safety.

6.CONCLUSIONS

Based on the results obtained under the conditions established and evaluated in this study, it can be concluded that:

- The lethal action of natural pesticides and entomopathogenic fungi on aphids in vitro showed promising results, with most of the formulations tested proving suitable for controlling the insects;
- Among the entomopathogenic fungi tested, the most effective species against aphids were Beauveria bassiana and Beauveria brongniartii with a TL_{50} of 5 and 6 days respectively;
- All the natural syrups tested had a mortality rate of over 80 per cent and a short TL_{50} of between 1 and 3 days.
- Based on the parameters of mean lethal time and mortality, the most efficient natural syrups were onion, garlic and chilli, neem, neutral detergent and cooking oil and soap and paraffin emulsion, resulting in the complete eradication of the aphid population in just one day;
- This study offers a more sustainable, safe and effective approach to managing pests that infest cultivated plants, bringing benefits to both farmers and the environment.

REFERENCES BIBLIOGRAPHIC

ABREU JUNIOR, H. de. Alternative Pest and Disease Control Practices in Agriculture. **EMOPI - Gráfica e Editora Ltda.** 1998. Campinas - SP.

AGROLINK - The Agricultural Content Portal. Check it out: Up-to-date news, weather forecasts, prices, specialist agricultural product systems, columnists, general classifieds. www.agrolink.com.br, 2024.

ALVES, S. B. Entomopathogenic fungi. **Microbial control of insects.** 2.ed. Piracacicaba: FEALQ, 1998. chap 11, p 289-382.

ANDERSON, R., BELL, A., BLANFORD, S., PAAIJMANS, K., & THOMAS, M. Comparative growth kinetics and virulence of four different isolates of entomopathogenic fungi in the house fly (Muscadomestica L.). **Journal of invertebrate pathology**, v.107 (3), p. 179-184, 2011

ASHOURI, S., & SHAYESTEH, N. Insecticidal activities of two powdered spices, black pepper and red pepper on adults of Rhyzopertha dominica (F.) and Sitophilus granaries (L.). **Munis Entomology and Zoology**, v. 5, p. 600-607, 2010

AYRES, M.I, ALVAREZ PUENTE, R.J et al. Primer for family producers: natural pesticides. **Editora INPA**, v. 1, p. 3- 34, 2020.

BARELLI, L., BEHIE, S., HU, S., & BIDOCHKA, M. Profiling Destruxin Synthesis by Specialist and Generalist Metarhizium Insect Pathogens during Coculture with Plants. **Applied and Environmental Microbiology**, v.88, e02474-21, 2022

BEZERRA, A. et al. "Efficiemcy of Essemtial Oils im the Comt ol of the Black Beam Aphid Aphis C accivo a Koch (Hemipte a: Aphididae)." **Brazilian Journal of Biology**, v. 83, p. 1678-4375, 2023.

BEZERRA, N. S., SOUSA, A. C. B. Efficiency of entomopathogenic fungi on leaf-cutting ants (hymenoptera: formicidae). **Dissertation (Master's)**, 2018.

BHOI, T., SAMAL, I., MAJHI, P., KOMAL, J., MAHANTA, D., PRADHAN, A., SAINI, V., RAJ, M., AHMAD, M., BEHERA, P., & ASHWINI, M. Insight into aphid mediated Potato Virus Y transmission: A molecular to bioinformatics prospective. **Frontiers in Microbiology**, v. 13, 2022.

BIANCALANI, T., & GORE, J. Disentangling bacterial invasiveness from lethality in an
expe imemtal host-pathogem system. **Molecular Systems Biology**, v.15, p. e8707, 2018.

BOTTI, J. M. et al. Alternative control of Brevicoryne brassicae (Hemiptera: Aphididae) with extracts from different plant species. **Revista Brasileira de Ciências Agrárias**, v. 10, p. 178-183, 2015.

BUSTAMANTE, D., OLIVA, M., LEIVA, S., MENDOZA, J., BOBADILLA, L., ANGULO, G., & CALDERON, M. Phylogeny and species delimitations in the entomopathogenic genus Beauveria (Hypocreales, Ascomycota), including the description of B. peruviensis sp. nov. **MycoKeys**, v. 58, p. 47 - 68, 2019.

CARRASCO, S. Sabor & Saber. Garlic: love it or hate it. p. 2. **Pontifical Catholic University of São Paulo** - PUC-SP, 2016

CAVALCANTI, M. F. Use of detergents as alternative pest control. **Revista de Ciências Agrárias,** v. 36, n. 2, p. 200-208, 2021.

CHUQUIBALA-CHECAN, B., TORRES, M. et al., In Vitro Biological Activity of Beauveria bassiana, Beauveria peruviensis, and Metarhizium sp. Against Hypothenemus hampei (Coleoptera: Curculionidae). **International Journal of Agronomy**, v. 2023, p.1-9, 2023

CICHOCKA, E., GOSZCZNSKI, W., et al. M. Chemical and physiological changes caused by aphids feeding on their host plants. **Polish Journal of Entomology,** v. 84, p. 233 - 248, 2015.

DAMPC, J., KULA-MAXIMENKO, M., MOLON, M., & DURAK, R. Enzymatic Defence
Response of Apple Aphid Aphis pomi to Increased Temperature. **Insects**, v. 11, p. 436, 2020.

DE HOOG,G.S.The genera Beauveria, Isaria, Tritirachium,and Acrodontium gen. nov. **Studies in Mycology**, v. 1, p. 1-41, 1972

DEVI, K. U., PADMAVATHI, J., et al. A study of host specificity in the entomopathogenic fungi Beauvaria bassiana (Hypocreales, Clavicipitaceae). **Biocontrol Science and Technology.** v. 18, p. 975-989, 2008.

DRIVER, F., MILNER, R.J., TRUEMAN, J.W.H. A taxonomic revision of Metarhizium based on a hylogenetic of rDNA sequence data. **Mycological Research**. v. 104, p.135-150, 2000.

ELWELL, H. E; MAAS, A - Natural pest and disease control- Zimbabwe, **Natural farming netsork**, 1995, 128 p.

FAN, J., XIE, Y., XUE, J. et al. Cellular Apoptosis of Fhemocytes from Dendroimus tabulaeformis Tsai et Liu Larvae Induced with the Secondary Metabollites of Beauvaria brongniartii. **Plos One,** v. 8 p- 71-600, 2013
FINNEY, D. J. Probit analysis. 3. ed. Cambridge: Cambridge **University Press**, 1971. FENIBO, E., IJOMA, G., NURMAHOMED, W., & MATAMBO, T. The Potential and Green
Chemistry Attributes of Biopesticides for Sustainable Agriculture. **Sustainability**, v. 14(21),p. 14417, 2022.

FERNÁNDEZ-BRAVO, M., GSCHWEND, F., MAYERHOFER, J., HUG, A., WIDMER, F.,& ENKERLI, J. Land-Use Type Drives Soil Population Structures of the Entomopathogenic Fungal Genus Metarhizium. Microorganisms, v.9, p. 1380, 2021.

FERNÁNDEZ-CONRADI, P., JACTEL, H., ROBIN, C., TACK, A., & CASTAGNEYROL,B. (2018). Fungi reduce preference and performance of insect herbivores on challenged plants. **Ecology**, v. 99 (2), p. 300-311, 2018.

FERREIRA, D. Sisvar: a Guide for its Bootstrap procedures in multiple comparisons. **Ciência e Agrotecnologia**, v. 38, p. 109-112, 2014.

FLINT, M. L. A. Aphids: integrated pest management for home, gardeners and landscape professional. Pest Notes, **University of California**. 4p. May 2000.

FONTES, E. M. G. et al. Biological control of agricultural pests. [Brasília, DF: **Embrapa**, 2020.

FRANCIS, F., JACQUEMYN, H., DELVIGNE, F., & LIEVENS, B. From Diverse Origins to
Specific Targets: Role of Microorganisms in Indirect Pest Biological Control. **Insects,** v.11, p. 533, 2020.

FREIRES, SAMYA THALYTA DOS SANTOS. The use of natural insecticides in agriculture: A review. **Course Conclusion Work (Agronomy),** v.1, p. 1-38, 2022.

GAMS, W., ROZSYPAL, J. Metarhizium flavoviride n. sp. Isolated from insects and soil**. Acta Botanica Neerlandica**. v. 22, p 518-521, 1973.

GARCÍA, M. A. G., GARCÍA, S. C.; GORDILLO, J. M. L.; MARTÍNEZ, R. F. M. Hongos entomopathogens as an alternative in Biological Control.

Kuxulkab', v. 15, n. 27, p.25-28, 2008.
GUERRA, M. S. Receituário caseiro: Alternatives for controlling pests and diseases of cultivated plants and their products. Brasília, **Embrater**, p. 166, 1985.

GUERRERO, J. R. Potencial del botón de oro (Tithonia diversifolia) como controlador biológico de hormiga arriera, Atta cephalotes (Hymenoptera: Myrmicinae). **Dissertation (Master's Degree in Biological Sciences)** - Universidad Del Valle, 2013.

HALDER, J., RAI, A.B., KONDARAM, M.H. Compatibility of Neem Oil and Different Entomopathogens for the Management of Major Vegetable Sucking Pests. **Natl. Acad. Sci. Lett**. v. 36, p.19-25, 2013.

HODGES, R. D., MARCHAND, E. N.; WRIGHT, D. J. The morphology of Metarhizium flavoviride and its role in the control of locusts and grasshoppers. **Mycological Research**, London, v. 101, p. 321-328, 1997.

HOLMES, K., BABENDREIER, D. et al. Introduction to the biopesticides manual. **CAB International**, p. 1- 159, 2018.

HOLTZ, F.G., SALOMÃO, S.L. et al. Efficiency of Citrus limon L. ripe peel extract in the control of Raoiella indica (HIRST, 1942) (Prostigmata: Tenuipalpidae). **Revista de Gestão Social e Ambiental**, v.18 (9), p. 1-16, 2023.

HUMBER, R. A. Entomopathogenic fungal identification: A systematic guide. **Mycological Research**, v. 97, n. 9, p. 1103-1109, 2003.

HUMBER, R. A. Fungal Pathogens of Insects. In: LACEY, L. A. (Ed.). Manual of Techniques in Insect Pathology. San Diego: **Academic Press**, 1997. p. 153-185.

ISLAM, W., ADNAN, M., SHABBIR, A., NAVEED, H., ABUBAKAR, Y., QASIM, M.,TAYYAB, M., NOMAN, A., NISAR, M., KHAN, K., & ALI, H. Insect-fungal-interactions:A detailed review on entomopathogenic fungi pathogenicity to combat insect pests... Microbial pathogenesis, v. 159, p.105-122, 2021.

IWANICKI, N., JÚNIOR, I., EILENBERG, J., & LICHT, H. Comparative RNAseq Analysis of the Insect-Pathogenic Fungus Metarhizium anisopliae Reveals Specific Transcriptome Signatures of Filamentous and Yeast-Like Development. **G3: Genes| Genomes | Genetics**, v.10, p. 2141 - 2157, 2020.

JARONSKI, S., & MASCARIN, G. Mass Production of Fungal Entomopathogens. **Academic Press - Microbial Control of Insect and Mite Pests**. v.n, p. 141-155, 2017.

JEFFS, L.B et al. Relationships between fungal spore morphologies and surface properties for entomopathogenic members of the genera Beauveria, Metarhizium, Paecilomyces, Tolypocladium and Verticillium. **Canadian Journal of Microbiology**, v. 45, p. 936-948, 1999.

JEON, H., KIM, K., KIM, Y., & LEE, S. Naturally occurring Piper plant amides potential in agricultural and pharmaceutical industries: perspectives of piperine and piperlongumine. **Applied Biological Chemistry**, v. 62, p. 1-7, 2019.

JIN, Y., WANG, Z., DONG, A., HUANG, Y., HAO, G., & SONG, B. Web repositories of
natural agents promote pests and pathogenic microbes management. **Briefings in bioinformatics**. v. 22 (6), p. 1-13, 2021.

KEPLER R. M., LUANGSA-ARD, J.J, HYWEL-JONES, N. L. A phylogenetically-based nomenclature for Cordycipitaceae (Hypocreales). **IMA Fungus** v. 8, p. 335-353, 2017.

KHOURY, C., GUILLOT, J., & NEMER, N. Lethal activity of beauvericin, a Beauveria bassiana mycotoxim, agaimst the two-spotted spider mites, Tet amychus u ticae Koch. **Journal of Applied Entomology**. v. 143, p. 974 - 983, 2019

KUMAR,S.,KUMAR,S. BHANDARI, D. &GAUTAM,M. Entomopathogens,Pathological Symptoms and their Role in Present Scenario of Agriculture: A Review. **International Journal of Current Microbiology and Applied Sciences,** v.9, p. 2110-2124, 2020.

LAGE DE MORAES, A., DE, R., PAES, A. Mycology. **Concepts and Methods for Training Professionals in Health Laboratories.** v. 3, p. 399-495, 2013.

LIMA, EVALDO R.S. Calda Sulfocálcica - control of the leprosis mite: other pests and diseases. Niterói**. EMATER - RIO**, 1993, 18p.

LIMA, M.L.F. Molecular characterisation of Metarhizium species and pathogenicity on Diatraea saccharalis. **Thesis (Doctorate)** - Federal University of Pernambuco, p. 1- 89, 2005.

LOPES, I. C. Conidia production of the entomopathogenic fungus Metarhizium anisopliae under different cultivation conditions and in a tray bioreactor. **Thesis (Master's) -** UNESP, 2016. 58 f.: il.

LOPES, JOÃO & RAMOS, ANDERSON & VILANOVA, EUCLIDES. Management of aphids and associated viruses in the safrinha maize crop. **Seminário Nacional de Milho Safrinha**, v. 21, p. 208-236, 2023.

MAHARANI, Y., MARYANA, N., RAUF, A., et al. Insect parasitoid and ant of associated on aphids (Aphididae) colonies on plants in West Java, v.3, p.59, 2020.

MAGALHÃES, B.P.; FRAZÃO, H.D. Effects of temperature, water content and substrate on conidial production of Metarhizium flavoviride **Revista de Microbiologia**, v.27, p.242-246, 1996.

MARGARITOPOULOS, J., et al. Long-term studies on the evolution of resistance of Myzus persicae (Hemiptera: Aphididae) to insecticides. **Greece. Bulletin of Entomological Research**, v. 111, p. 1 - 16, 2020.

MANNINO, M., HUARTE-BONNET, C., DAVYT-COLO, B., & PEDRINI, N. Is the Insect Cuticle the only Entry Gate for Fungal Infection? Insights into Alternative Modes of Action Entomopathogenic Fungi. **Journal of Fungi**. v.5 (2), p. 33, 2019.

MASCARIN, G., LOPES, R., DELALIBERA, Í., FERNANDES, É., LUZ, C., & FARIA, M.
Current status and perspectives of fungal entomopathogens used for microbial control of arthropod pests in Brazil **Journal of invertebrate pathology,** v. 165, p. 46-53, 2019.

MEENA, R., KUMAR, S., DATTA, R., LAL, R., VIJAYAKUMAR, V. et al. Impact of Agrochemicals on Soil Microbiota and Management: A Review. **Restoring Degraded Lands to Meet UN-SDGs,** v. 9, p. 34, 2020.

MENEZES, M. S. D et al. Extracts from the algarobeira tree as an alternative for controlling the black bean aphid Aphi craccivora. **Open Journal Systems**, v.1, p. 47-53, 2019.

MESQUITA, E., HU, S., LIMA, T., GÔLO, P., & BIDOCHKA, M. Utilisation of Metarhizium as an insect biocontrol agent and a plant bioinoculant with special reference to Brazil. **Frontiers in Fungal Biology**, v. 4, p. 103, 2021.

MICHEREFF FILHO, M.; LIMA, M. F. Management of whitefly, geminiviruses and criniviruses in potato cultivation. **Technical Communication**, p. 12, 2016.

MOURA, M. R.; SANTOS, C. C. Effects of soap and detergent on insect mortality.
Brazilian Journal of Entomology, v. 62, n. 4, p. 345-352, 2018.

NOMURA, M.; SANTOS, A. M.; ROCHA, A. F.. Control of whitefly (Bemisa ssp.) on tomato seedlings with plant extracts**. Ciência Et Praxis,** v. 25, p. 602-606, 2018.

NONG, X., LIU, C., LU, X., WANG, Q., WANG, G., & ZHANG, Z. Laboratory evaluation entomopathogenic fungi against the white grubs, Holotrichia oblita and Anomala corpulenta (Coleoptera: Scarabaeidae) from the field of peanut, Arachis hypogaea. **Biocontrol Science and Technology**, v. 21, p. 593 - 603, 2011.

OLIVEIRA, R. F.; CAVALCANTE, H. S. Insecticidal properties of capsaicin in insects. **Journal Pesticide Science**, v. 42, n. 1, p. 23-30, 2017.

ORTIZ-URQUIZA, A., & KEYHANI, N. Action on the Surface: Entomopathogenic Fungi versus the Insect Cuticle. **Insects**, v.4, p. 357 - 374, 2013.

OZDEMIR,I.,TUNCER,C.,ERPER,I.,& KUSHIYEV,R.. Efficacyof the entomopathogenic fungi; Beauveria bassiana and Metarhizium anisopliae against the cowpea weevil, Callosobruchus maculatus F. (Coleoptera: Chrysomelidae: Bruchinae). **Egyptian Journal of Biological Pest Control**, p. 30, 2020.

PARRA, J. R. P.; COELHO JUNIOR, A. Applied Biological Control in Brazil: From Laboratory Assays to Field Application. **Journal of Insect Science**, v. 19, p.1-6, 2019.

PASTOR, A. R. Control of castaño (Castanea sativa) carpophagous insects in Andalusia through mass capture with sex pheromone and evaluation of the insecticidal activity of entomopathogenic weeds. **Thesis (Doctorate)** - Universidad de Córdoba, Córdoba, v. 42013.

PEDRINI, N. Molecular interactions between entomopathogenic fungi (Hypocreales) and their insect host: Perspectives from stressful cuticle and haemolymph battlefields and the potential of dual RNA sequencing for future studies. **Fungal Biol.**, v. 122, p. 538-545, 2018

PEDRINI N., ORTIZ-URQUIZA A., HUARTE-BONNET C., ZHANG S, KEYHANI N. Targeting of insect epicuticular lipids by the entomopathogenic fungus Beauveria bassiana: hydrocarbon oxidation within the context of a host-pathogen interaction. **Front Microbiol**, v. 4, p. 24, 2013.

PENTEADO, SILVIO R., Bordeaux mixture - How and why to use it. Cati/Dextru - Campinas, SP. **Technical Leaflet**, 1996, 6p.

PORRAS, M. AGUDELO-CANTEROG.SANTIAGO-MARTÍNEZ et al. Fungal infections lead to shifts in thermal tolerance and voluntary exposure to extreme temperatures in both prey and predator insects. **Scientific Reports**, p. 11, 2021.

RIBA, G.; BOUVIER-FOURCADE, I.; CAUDAL, A. Isoenzymes polymorphism in Metarhizium anisopliae (Deteromycotina: Hyphomycetes) entomogenous fungi. **Mycopathologia,** v. 96, p.161-169, 1986.

ROUF, R., UDDIN, S., SARKER, D. et al. Antiviral potential of garlic (Allium sativum) and its organosulfur compounds: A systematic update of pre-clinical and clinical data. **Trends in Food Science & Technology**, v. 104, p. 219 - 234, 2020

RUMBOS, C., & ATHANASSIOU, C. Use of entomopathogenic fungi for the control of stored-product insects: can fungi protect durable commodities? **Journal of Pest Science**, v. 90, p. 839-854, 2017.

SANTI, L., SILVA, W., PINTO, A., SCHRANK, A., & VAINSTEIN, M. (2010). Metarhizium anisopliae host-pathogen interaction: differential immunoproteomics reveals proteins involved in the infection process of arthropods... Fungal biology, v. 114 (4), p. 312- 319

SANTOS, A., ZANETTI, R.; ZANUNCIO, J. C.; MENDONÇA, L. A.; SOUZA-SILVA, A.;
MEDEIROS, A. G. B. Efficiency and cost of controlling Atta spp. (Hymenoptera: Formicidae) nests with thermonebulisation. **Cerne,** v. 13, p. 23-27, 2007.

SIMON, J., & PECCOUD, J. Rapid evolution of aphid pests in agricultural environments. **Current opinion in insect science**, v. 26, p. 17-24, 2018.

SOFWAH, A., & PRASTOWO, S. Efektivitas Beauvaria bassiana dan Metarhizium anisopliae terhadap Pengendalian Hama Thrips sp. (Thysanoptera:

Tripidae) pada Tanaman Cabai Besar (Capsicum annuum L.). **Berkala Ilmiah Pertanian**, 2023

SOUZA, F. J., SOUSA, A.C.B. Coadunability between Aspergillus sp. NA01 with insecticides used in the integrated management of the rust caterpillar Agrotis ípsilon. **Nova Edições Acadêmicas**, v (n), p. 1-53, 2021.

SOUZA, M. F. et al. Evaluation of the repellent potential of onion extracts against insects. **Revista de Ciências Agrárias,** v. 35, n. 2, p. 145-152, 2020.

SPARKS, T., HAHN, D., & GARIZI, N. Natural products, their derivatives, mimics and synthetic equivalents: role in agrochemical discovery. **Pest Management Science**, v.73 (4), p. 700-715, 2017.

SPAZZIN, M. P., CEMIN, M. F. R..; BELLON, P. P. . Insecticidal potential of agroecological sprays in the control of cabbage curuquerê, Ascia monuste orseis (Lepidoptera: Pieridae). **Research, Society and Development**, v. 12, n. 4, p. e1812440815, 2023.

TÉLLEZ-JURADO, A.; RAMÍREZ, M. G. C.; FLORES, Y. M.; TORRES, A. A.;ARANACUENCA, A. Mecanismos de acción y respuesta en la relación de hongos entomopatógenos y insectos. **Revista Mexicana de Micología**, v. 30, p.73-80, 2009.

TENNAKOON, D. Life in leaf litter: Fungal community succession during decomposition. **Mycosphere,** v. 12 p. 406-429, 2021

TIAGO, P., FUNGARO, M., & FURLANETO, M. Cuticle-deg adimg p oteases f om the entomopathogen Metarhizium flavoviride and their distribution in secreted and intracellular fractions. **Letters in Applied Microbiology,** v.34, p. 91-94, 2002.

TRÊS, F., RESENDE, S. - Supermagro, enriched biofertiliser. Niterói, **EMATER- RIO,**1995, 11p.

TUDI, M., RUAN, H., et al. Agriculture Development, Pesticide Application and Its impact on the environment. **International Journal Environmental Research and Public Health,** v.18, p. 1112, 2021.

TULLOCH, A.P. Chemistry of waxes of higher plants. in: kolattukudy, p. (ed.), **Chemistry and biochemistry of natural waxes**, ed, p. 236-252, 1976.

UKABHAI, V., & MF, A. Bionomics of aphid, Aphis gossypii (Glover) on cumin, Cuminum cyminum Linn. under laboratory conditions. **Journal of**

Entomology and Zoology Studies, 2022.

VIVEKANANDHAN, P., SWATHY, K., ALFORD, L. et al. Toxicity of Metarhizium flavoviride conidia virulence against Spodoptera litura (Lepidoptera: Noctuidae) and its impact on physiological and biochemical activities. **Sci Rep**, v. 12, 16775, 2022.

WANG, C., & FENG, M. Advances in fundamental and applied studies in China of fungal biocontrol agents for use against arthropod pests. **Biological Control**, v. 68, p. 129-135, 2014

WANG, M.C., GONG, M., ZANG, H.B. et al. Effect of Methamidophos and Urea Application on Microbial Communities in Soils as Determined by Microbial Biomass and Community Level Physiological Profiles. **J. Environ. Sci. Health B**, v. 41, p. 399-413, 2006

WOJCIECHOWSKA, M., STEPNOWSKI, P., & GOLEWBIOWSKI, M. The use of insecticides to control insect pests. **ISJ-Invertebrate Survival Journal**, v. 13, p. 210-220, 2016.

XIAO, G., YING, S., ZHENG, P. et al. Genomic perspectives on the evolution of fungal entomopathogenicity in Beauveria bassiana. **Scientific Reports,** v. 2, no. 1, p. 483, 2012.

ZIMMERMANN, G. Review on safety of the entomopathogenic fungi Beauveria bassiana and Beauveria brongniartii. **Biocontrol Science and Technology**, v. 17, n. 5-6, p. 553-596, 2007.

ZIABARI, O., LI, B., HARDY, N., & BRISSON, J. Aphid male wing polymorphisms are transient and have evolved repeatedly. **Evolution; international journal of organic Evolution,** v. 77, p. 1056-1065, 2023.

MIX
Papier aus verantwortungsvollen Quellen
Paper from responsible sources
FSC® C105338

Printed by Books on Demand GmbH, Norderstedt / Germany